Ana Maria Jesus Sousa-Cunha

Intra-puparial Development in Sarcophagidae (Diptera)

Ana Maria Jesus Sousa-Cunha

Intra-puparial Development in Sarcophagidae (Diptera)

LAP LAMBERT Academic Publishing

Impressum / Imprint
Bibliografische Information der Deutschen Nationalbibliothek: Die Deutsche
Nationalbibliothek verzeichnet diese Publikation in der Deutschen
Nationalbibliografie; detaillierte bibliografische Daten sind im Internet über
http://dnb.d-nb.de abrufbar.
Alle in diesem Buch genannten Marken und Produktnamen unterliegen
warenzeichen-, marken- oder patentrechtlichem Schutz bzw. sind
Warenzeichen oder eingetragene Warenzeichen der jeweiligen Inhaber. Die
Wiedergabe von Marken, Produktnamen, Gebrauchsnamen, Handelsnamen,
Warenbezeichnungen u.s.w. in diesem Werk berechtigt auch ohne besondere
Kennzeichnung nicht zu der Annahme, dass solche Namen im Sinne der
Warenzeichen- und Markenschutzgesetzgebung als frei zu betrachten wären
und daher von jedermann benutzt werden dürften.

Bibliographic information published by the Deutsche Nationalbibliothek: The
Deutsche Nationalbibliothek lists this publication in the Deutsche
Nationalbibliografie; detailed bibliographic data are available in the Internet
at http://dnb.d-nb.de.
Any brand names and product names mentioned in this book are subject to
trademark, brand or patent protection and are trademarks or registered
trademarks of their respective holders. The use of brand names, product
names, common names, trade names, product descriptions etc. even without
a particular marking in this work is in no way to be construed to mean that
such names may be regarded as unrestricted in respect of trademark and
brand protection legislation and could thus be used by anyone.

Coverbild / Cover image: www.ingimage.com

Verlag / Publisher:
LAP LAMBERT Academic Publishing
ist ein Imprint der / is a trademark of
OmniScriptum GmbH & Co. KG
Heinrich-Böcking-Str. 6-8, 66121 Saarbrücken, Deutschland / Germany
Email: info@lap-publishing.com

Herstellung: siehe letzte Seite /
Printed at: see last page
ISBN: 978-3-659-74295-8

Zugl. / Approved by: Brasilia, University of Brasilia, Diss., 2014

"... Behold, I will send swarms of flies on you ... and also the ground whereon they are."

Exodus 8:21

Marcial Fernandes, my husband and my parents Ozanira Maria de Jesus Sousa (*in memoriam*) and Francisco Sousa Firmiano, my eternal gratitude.

ACKNOWLEDGEMENTS

God who has always been and will continue to be the reason for everything in my life and for giving me the strength and the courage needed to achieve this work.

My husband, understanding that stood by me, for the many times that accompanied me in the experimental activities, sleeping without a minimum of comfort in the laboratory.

My Sousa family for understanding my absence and for always believing in my dreams.

Rodrigo Meneses de Barros, for sharing your knowledge Sarcophagidae.

Dr. Júlio Mendes and Dra. Patrícia Jacqueline Thyssen for their suggestions and contributions in the qualifying examination.

Dr. Gino Chaves da Rocha and Dra. Ivone Rezende Diniz for accepting the invitation to compose the examining board, in addition to valuable suggestions during the defense of the dissertation.

Ana Carolina Franco, the affection of her friendship, the strength in times of weakness and encouragement in this task.

All teachers: Rosana Tidon, John Ray, Raul Laumann, Miguel Borges, Marisa Frizzas, Rodrigo Gurgel, Ivone Diniz, which significantly contributed to my academic journey.

All colleagues from the laboratory of forensic entomology at UNB, for everything.

The staff of UNB, which contributed even indirectly, to carry out this work.

SEEDF for granting clearances for participation at conferences in Germany and Brazil.

Graduate Program in Zoology, FINATEC for financial support for the international congress and the funding agencies: CAPES, CNPq.

SUMMARY

1. INTRODUCTION

The Order Diptera is among the four megadiverse insect orders and gathers mosquitoes and flies, comprising about 160,000 described species, (Pape e*t al,* 2011; Brown *et al,* 2009). In the Neotropical region are known more than 31,000 species in 118 families (Carvalho *et al,* 2012). Sarcophagidae Family is generally composed of robust individuals of gray and varied size. Exhibit chest with three sharp black stripes and abdomen in checkerboard pattern, striped or spotted (Shewell 1987; Pape & Dahlem 2010; Oliveira-Costa *et al,* 2011).

The flesh flies are present in almost everyone with more than 3,000 described species in nearly 180 genres (Pape *et al,* 2011). Presents a very varied biology, species may have habits dung, carrion, some are parasitoids, predators of other arthropods also parasitizing reptiles, amphibians and mollusks. Others are considered mechanical vectors of pathogens or, myiasis-causing in vertebrates (Zumpt 1965; Guimarães & Papavero 1999). Most flesh flies larvae are scavenger feeding on decaying organic matter, giving these species forensic importance (Ishijima 1967; Barros *et al,* 2008; Buenaventura *et al,* 2009).

The larvae present in a subsequent spiracles of each cavity being used for identifying characteristics of this group. The mandible is usually strong and the pharyngeal head skeleton is great. The peritreme is incomplete both the second and the third instar and has no scar ecdisial (Shewell 1987; Pape & Dahlem 2010).

Peckia intermutans Walker, 1861 is a sarcophagid with geographical distribution Nearctic and Neotropical (Buenaventura & Pape 2013). Their larvae are scavengers with multiple records of its development in animal carcasses, such as snakes and small rodents (Moretti *et al,* 2008; Ledo *et al,* 2012). Previous studies have considered it as a potential forensic importance not only for being among the most common forensic species found in organic animal matter in decomposition, but have been found in human cadavers (Carvalho & Linhares 2001; Carvalho and Mello-Patiu 2008; Buenaventura *et al*, 2009;. Vairo *et al*, 2011; Oliveira-Costa *et al*, 2011).

Bionomic aspects of *P. intermutans* as life expectancy and fertility were analyzed by Oliveira et al. (2002b). 1st instar larvae were described with emphasis on the importance of clypeal arch and jaw as features that aid in Sarcophagidae classification (Lopes 1982), and the 3rd instar were described with the aid of scanning electron microscopy, where the following were analyzed characters: the cephalic segment, the mouth hooks, the anterior and posterior spiracles, tubers and spiracular rays (Jirón & Bolaños 1986). The post-embryonic development was evaluated on different diets, which was demonstrated by preference larvae diet containing meat, since under natural conditions they develop animal carcasses (Loureiro *et al.*, 2005).

Peckia lambens Wiedemann, 1830 as other species of flesh flies, it is considered vector potential pathogens such as viruses, bacteria, protozoa and helminthes, easily collected from human feces (Marchiori *et al*, 2003). There are records of the species and as many parasite or parasitoid and host of other animals (Rocha & Mendes 1996; Marchiori *et al*, 2003, 2007; Hernández *et al*, 2009; Silva *et al*, 2012). The species has been described as *Sarcodexia sternodontis* was registered as optional parasite of various arthropods, being created from dead aquatic snails *Marisa cornuarietis* found on the bank of a canal in Miami in 1971 (Stegmaier 1972). To investigate the frequency, abundance and seasonality of Sarcophagidae in Rio de Janeiro Zoo it was found that *P. lambens* was more abundant during the winter months (Oliveira *et al*, 2002a).

P. lambens it is also causing myiasis and one of the reported cases occurred in *Epipedobates trivittatus* specie of poisonous frog (Hagman *et al*, 2005). In the state of Goias, Brazil, was the first recorded case of human myiasis often 12.1% of reported cases. This percentage placed her second as a cause of secondary screwworm infestation in this study (Fernandes *et al*, 2009).

According Ledo et al. (2012) there are not many reports in the literature about decomposition rate of other terrestrial vertebrates, particularly amphibians and reptiles, especially decomposition of small carcasses. However, the authors reported some carrion insects associated with these carcasses and provided information on the colonization and Sarcophagidae establishment in small housings. *P. lambens* was found in carcass of *Rhinella schneideri*. Oliveira-Costa *et al.* (2011) already collected the species colonizing human corpses, both adults and immature.

Other studies in egg description area of the larval instars and puparia were also made and brought important contributions to the knowledge of the species. Leite & Lopes (1989), when they presented the differences between the first instar *P. lambens* and *P. chrysostoma* after describing them separately. Lopes & Leite (1989), described the egg with the aid of scanning electron microscopy, this has cylindrical shape is long and presents tapering at the ends. It has dorsal curved surface, ventral and laterally. Although viviparity the species, its egg has an intra chorionic mesh with characteristics similar to ovoviviparous species Calliphoridae.

The three larval instars of *P. lambens* were described with the aid of optical microscopy and scanning electron microscopy. The observed and analyzed aspects were the average length, the color, the skeleton head, thoracic and abdominal segments, tubers, processes and previous spiracles. Has been made yet, a description of the puparium which is similar to that of other flies. It is cylindrical and the housing is rigid and darkened. With the aid of scanning electron microscopy, we observed that the thorns bands are similar, being tapered in the apical part and the basal part, full-bodied. There are not apparently structures that differ this species from another, since there are few studies that bring a full discussion of the stages of its development (Vairo 2011).

1.1 TERMINOLOGY USED IN THE STUDY ON INTRA-PUPARIAL DEVELOPMENT

P. intermutans and *P. lambens* are holometabolous insects that during development, they undergo structural, physiological and anatomical changes. Fraenkel & Bhaskaran (1973) presented concepts and terminology applied to intra-puparial development and metamorphosis, though some of the concepts as apolysis and pharate had already been discussed enough in Hinton (1946, 1971 and 1973), Jenkin & Hinton (1966) Wigglesworth (1973) and the pre-pupae concept discussed by Costa & Vanin (1985).

Intra-puparial development involves two processes: (1) the pupariation has been described as a complex process of morphological and structural changes (Fraenkel & Bhaskaran 1973; Delinger & Zdárek 1994). From that moment, pupae are inside a barrel-shaped structure which is formed by the cuticle of the last larval instar called puparium. This is formed from the retraction of the first three larval segments into the body followed by a shortening of approximately ¾ of the length of the larvae. And (2) process called, itself, pupation, where all the morphological, structural and anatomical changes will occur in the pupa until the time of adult emergence from within the puparium.

The terminology used in this work to describe the morphological changes that occurred during the phases of intra-puparial development *P. intermutans* and *P. lambens*, and a description of pupariation and pupation process followed the one proposed by Fraenkel & Bhaskaran (1973), by Cepeda-Palacios & Scholl (2000) and reviewed by Barros-Cordeiro et al. (2014).

Pupariation – It corresponds to the period in which the mature larva ceases feeding and completes its immobilization and reduction of its size and the larval cuticle pigments up and hardens.

Larval-pupal apolysis – is the process which results in the formation of adult epidermis and their subsequent separation from the last larval cuticle which form the puparium, after completion of pupariação process.

Cryptocephalic Pupa – is known as the phase of head hidden, it is impossible to externally observe the same distinction and thoracic appendages.

Phanerocephalic Pupa – extraversion head and thoracic appendages is complete, and is the stage where begins the process of apólise between pupa and adult.

Pharate adult – It is the longest phase of the intra-pupal development extending from the end of apólise pupa-adult to adult emergence corresponding to adult maturation.

Imago – corresponds to the adult formed after metamorphosis.

1.2 LITERATURE REVIEW OF THE DESCRIPTION OF IMMATURE SARCOPHAGIDAE, SARCOPHAGINAE

Sarcophagidae is a diverse family in the order Diptera and is present in all biogeographic areas of the world with approximately 3,000 described species in almost 200 genera (Shewell 1987; Pape et al, 2011.). It is divided into three subfamilies (Miltogramminae, Paramacronychiinae and

Sarcophaginae), with the latter dominating the fauna of the Neotropical region (Pape & Dahlem 2010).

The flesh flies present a varied biology, species may have habits dung, carrion, some are parasitoids or predators of other insects, parasites and some snails and earthworms. Others are considered mechanical vectors of pathogens or, myiasis-causing in vertebrates (Zumpt 1965; Guimarães & Papavero 1999).

Females are viviparous or ovoviviparous depositing 1st instar. These larvae have some peculiar characteristics as later spiracles in cavity large cephalopharyngeal skeleton and generally strong jaws (Shewell 1987). Most flesh flies larvae is scavenger feeding on decaying organic matter, giving the these species forensic importance (Ishijima 1967; Barros *et al*, 2008; Buenaventura *et al*, 2009).

1.2.1 *Descriptive studies of the immature stages realized outside Brazil*

Flesh flies larvae are easily identified at the family level, however, interspecific and have very similar characteristics subgenerically (Aspoas 1991). Identification and descriptive study of the immature stages is essential for knowledge of the species. Larvae have numerous characters useful to identify its specie. Wherein the cephalopharyngeal skeleton, the most useful, because it shows morphological variations within the group. However, there are other features which are widely studied for distinguishing species and instar (Knipling 1936).

The size and overall appearance of the larvae characteristics are limited, once the larvae to feed its size increase rapidly. It must be observed and analyzed immediately after larviposition. The shape, size and distribution of the spines are considered as one of the most reliable in the differentiation of species as there are wide variations. The rear cavity has some thorns on its edge and small tubers located both above and below the rim. The size of the cavity, the arrangement of thorns and the amount of tubers are important characteristics used to identify the species. Finally, the anal area where a differentiation also occurs in the shape and size. There is also the presence of tubers can vary in amount and form (Knipling 1936).

They have a cylindrical body with front end and rear tapered truncated. They lack cephalization but present head skeleton. The oral appliance is reduced to hooks or jaws which move in a vertical plane. The twelve body segments are well defined presenting thorns bands that vary in shape and size and may have lots of these in the ventral region. The last abdominal segment has tubers which are located around the spiracular plate and usually twelve in number (Fontoura *et al,* 2013).

The anterior spiracles have short branches and are located at the side of the second segment. The posterior spiracles are in the last abdominal segment in a spiracular cavity, which is one important feature that differentiates from other family Sarcophagidae (Shewell 1987; Fontoura *et al*, 2013). Another distinguishable characteristic is the absence of the button on the plate spiracular (Greene 1925). The posterior spiracles also present variations in shape and size. For example, the species that are scavengers or parasites, the spiracular slits do not exhibit peritreme. Cantrell (1981) emphasized that attention should be paid to the number of rays in the previous spiracles, the shape of the posterior peritrema spiracles and cephalopharyngeal skeleton. Also, the disposition and the shape of the slits in posterior spiracles and the presence of an ecdisial scar, they are useful tools in the identification of species of Sarcophagidae. Breathing is amphipneustic by presenting a couple of spiracles which can be located in the dorsolateral or side prothoracic segment. They have finger-like branches in the short distal region and may have variation in the amount and arrangement in different species (Guimarães & Amorim 2006).

In Sarcophagidae, peritreme of the posterior spiracle is divided into four arcs. The inner ends of the bows are of two types: one is conical or without a turgid end and the other end with a turgid. Previous bows are straight or slightly curved (Ishijima 1967).

Some important observations on the *Sarcophaga* gender were registered in the studies. There are at least two distinct types of posterior spiracles (Root 1923). Among the species there are huge variations in the first instar (Knipling 1936), and yet, some larvae may be optional parasites for a certain period or until they complete their development (Zumpt 1965).

There scavenger species, saprophagous, coprophagous that present preference for carnivores droppings, however, there are others that are not particularly attracted by carcasses or excrement (Banks 1912; Bohart & Gressitt 1951).

1.2.2. *Descriptive studies of the immature stages carried out in Brazil*

In Brazil, most of the known descriptions were carried out by Professor Dr. Hugo de Souza Lopes, who stressed the importance of studying this family by the small number of studies that had until then. Also reported difficulty was in identifying the species, since females are relatively similar, and males are mainly identified by the male genitalia.

In his 1943 work, Lopes contributed to the knowledge of the larvae of this family presenting important features of the head skeleton of eleven species compared to *Musca domestica*. One of the features highlighted by the author which differs from the group of *M. domestica* is the presence of an incomplete ring which is situated on the ventral part of early in the gut and after speaking.

Other characters were analyzed: the head skeleton has quite chitin, the great food reservoir; the first segment generally has spikes on the leading edge are varied in form and in position and can gather a pigmented board that also varies according to species. The sclerite lip is formed by a pair of symmetrical parts which are joined generally in a greater or lesser extent in a wide base (Lopes 1943).

Some unique features were also observed and recorded as, for example, larvae instar 1 *Titanogrypa (Cucullomyia) larvicida* having poison capable of preventing the larvae of other species living in the same nutrient medium and the first instar of this kind has a duration five days (Lopes 1935).

Another specie, *Titanogrypa luculenta* also has poison and has habit predator of other larvae as T. larvicide. And both have protection from the poison of the larvae of the same species. Another specific character *T. luculenta* is that by their larvae have spread over the abdominal segments (Lopes 1976).

In the 1945 work, Lopes noted that the larvae of 1st instar genre *Notochaeta* Aldrich, 1916 did not have the dorsopharyngeal sclerite and are obligatorily parasites both invertebrates and vertebrates (Lopes & Vogelsang 1953). The 1st stage larvae of the species *N. confusa* and N. *aldrichi* lasts less than 24 hours and about 70 hours after the larvae penetrate the host, the worm pupariation it initiate the process. According author, the rapid decomposition of the *Oligochaeta* justify this short larval period, which is half the time required for the larvae of other flesh flies.

The specie *Tricharaea (Sarcophagula) canuta*, the larvae showed ornamentation in pseudocephalon, and the pharyngeal skeleton similar to that observed in *Sarothromyia* and *Oxysarcodexia* genres. And posterior stigmas are not within a cavity (Lopes 1954).

According to the observations made by Kano & Lopes (1971), the larva of *Sarcophaga (Johnsonimima) aurora* performed well individualized jaws feature found only in Palaearctic species such as *Sarcophaga carnaria*. In *Retrocitomyia trinitatensis* specie was found a similar jaw to those present, found in species of Raviniini tribe. However, the type-described species of *Retrocitomyia* gender were not found traces jaw (Lopes 1985).

Another relevant study was on the Sarcophagidae classification, pointing to the importance of using the development of clypeal bow and 1st larvae jaw urge for species identification (Lopes 1982).

According to Lopes and Leite (1986) the larvae of 1st instar of Raviniini tribe have a clear synapomorphy which is a fluted strip (festoons) next to pseudocephalic grooves, so considering a monophyletic group. In 1975, Lopes had proposed a subtribe, Oxysarcodexiina, for species that show this effect. Another synapomorphy found was the presence of small teeth in the larvae of *Oxysarcodexia* genre, which varies in amount depending on the species.

1.2.3 *Studies pupae and pupae of Sarcophagidae*

Studies describing pupae and pupae are incipient and mostly report some external characteristics that can help identify the species (Zumpt 1965), and others bring a small reference pupal period total species (Lopes 1945).

A dichotomous key to identify pupae of 41 species of flesh flies was made a detailed description of each puparium. The cavity present in the back end was later named this work as well and was also reported the difficulty in identifying the species without making a cut in this cavity, since spiracular plates which are inserted into the same feature peculiar characteristics for each species (Greene 1925).

Another dichotomous key was written to identify both the pupae of flesh flies as other families surveyed in the study. The analyzed characteristics were micro tuber present on the surface of the puparium, distribution and arrangement. The posterior spiracles that according to the observations were similar to the larva and the subsequent cavity *S. gressitti* were shallower than in *S. ruficornis* (Bohart & Gressit 1951).

Newhouse *et al.* (1955) described the pupae of three species of the genus *Sarcophaga*, concluding that there is a lot of similarity between them. The criteria were that the puparium presented elliptical and very dark red color way. The opening of the cavity was spiracular oval to elliptical. The spiracular plate was a reddish-brown deep contrast to the almost white cracks. Tubers around the posterior cavity were flattened and distorted. The annals tubers were prominent. The spines were complete bands between the segments 3 to 12. Prothoracic blowholes evident, however, it was not possible to discern the number of digits. Among the three species the authors failed to see distinction in their pupae. However, in *S. shermani* the bridge that connected the annals tubers and subsequent tuber was generally poorly developed or even absent. According to study of the puparium of the genre *Helicobi*a noted that it were similar to the *Sarcophaga*, except for size (Fuller 1938).

The characteristics of the last larval instar may assist in identifying the species, since they are retained when the pupa is formed by the last larval cuticle. Anal plate can also be used for the diagnosis of species, usually keeps it out. Was made a description of the pupae of *Parasarcophaga knabi* and *Tricharaea brevicornis* with the aid of light microscopy and a difference was observed, the *T. brevicornis* pupae did not have this cavity in the rear end as *P. knabi* and other flesh flies and presents a flattening in the region (Ferrar 1979).

In scanning electron microscopy studies conducted with seven flesh flies occurring in Australia, it was observed that pupae were similar to each other. Larval spiracular posterior slot has been preserved, but the aperture size varied so that it became difficult to see spiracular plates in some species. Another important observation was that there are no visible present respiratory structures in pupae (Ferrar 1979; Cantrell 1981).

Some works bring a comparative description of the pupae from different families in order to increase knowledge on the morphology of the pupae for future applications in the field of forensic entomology, emphasizing the importance of scanning electron microscopy which showed details that can be used as evidence entomological (Sukontason *et al,* 2006; Singh *et al,* 2012).

Studies in Brazil with the aid of optical microscopy described the morphological characteristics of the buds located on the back end of the stigmata, the concavity present in the annals stigmas and tubers of several species, such as the *Emblemasoma* genre that has pupae with small buds , chamber of very deep stigma later with a small opening (Lopes 1971). In *Dexosarcophaga*, the papillae at the posterior end, are long, the stigmas are in a shallow concavity and wide open (Lopes 1946, 1968). However in *Tricharaea* gender was not found the back cavity and as noted, the spiracles were external. And *Sarcophaga*, the puparium performed well developed ventral buds and the camera spiracular pretty deep (Lopes 1983).

According to Lopes (1982) the puparium of the species *Sarconeiva fimbriata* presented a peculiar characteristic that differed from other pupae of flesh flies, previous stigmas were well developed, and one explanation for this variation would be an obvious adaptation for living in small gastropods.

With scanning electron microscopy it was possible to observe in detail the characters that are not visible by optical microscopy. Another important point is that the puparium of flesh flies show almost all the features observed in the third instar larva, with one exception, the head region is fully retracted at this stage (Mendonça et al, 2013).

Regarding the pupa time several records that vary according to temperature, humidity and photoperiod (Saloña Bordas & González-Mora 2005). However, the differences are very small, but persistent in the length of the life cycle, clearly the pupal period (Aspoas 1991). The pupal time recorded for the genus Sarcophaga the average was 12.2 days at temperatures between 23 ° C and 26 ° C, relative humidity 50% ± 10% (Kamal 1958; Aspoas 1991; Saloña Bordas & González-Mora 2005; Draber-Monko et al., 2009).

Studies in several species of the genus Peckia Robineau-Desvoidy, 1830 presented the following results for the length of pupal period: P. chrysostoma was 23.5 days ± 1.3 days at 18 ° C; 12.5 ± 0.7 days at 27 ° C and 15.0 ± 0.7 days at room temperature. To the obtained P. ingens pupa was 33.0 ± 2.2 days at 18 ° C; 16.0 ± 1.0 days at 27 ° C and 19.0 ± 1.0 days at room temperature (Ferraz 1995). The pupal period P. gulo was 8-11 days (Méndez & Pape 2002). The pupa time P. smarti during the dry season was 6.27 days and 6.39 days P. pallidipilosa was. In the rainy season it was 9.73 days and 8.95 days, respectively (Oliveira-da-Silva et al., 2006).

This work aimed to describe the morphological changes during the development of P. intermutans and P. lambens under different controlled conditions of laboratory and analyze larval and pupal viability and sex ratio in the three temperatures used in the experiment with the second species.

2. INTRA-PUPARIAL DEVELOPMENT IN *PECKIA INTERMUTANS* AND *PECKIA LAMBENS* (DIPTERA, SARCOPHAGIDAE)

1. INTRODUCTION

Entomological forensic evidence has shown great and decisive significance in post-mortem investigations, once it are found at crime scenes, larvae and pupae particularly dipterous of the families: Calliphoridae, Muscidae, Sarcophagidae, among others. In addition to establishing in many cases the differences between the scene of the crime and the place where the body was found, the time elapsed between the death and the availability of the corpse to colonization by insects on an estimation of postmortem interval (Carvalho *et al*, 2000; Ames *et al*, 2006; Pujol-Luz *et al*, 2006; Oliveira-Costa *et al*, 2011; Miranda *et al*, 2013).

The flesh flies have been found during the decomposition process carcasses. Like most species are viviparous or ovoviviparous, it is believed that there may be a precursor exploitation of this food source for the larvae deposited there (Romera *et al*, 2003; Barros *et al*, 2008; Vairo *et al*, 2011; Ledo *et al*, 2012; Mello-Patiu *et al*, 2014).

Peckia intermutans Walker, 1861, is a sarcophagid with geographical distribution Nearctic and Neotropical (Buenaventura & Pape 2013). Its larvae were considered by scavenger develop animal carcasses (Loureiro *et al*, 2005; Barros *et al*, 2008; Moretti *et al*, 2008; Barbosa *et al*, 2009; Rosa *et al*, 2009; Ledo *et al*, 2012). It was appointed as an indicator of forensic importance (Carvalho & Mello-Patiu 2008; Buenaventura *et al*, 2009; Vairo *et al*, 2011), with record on human cadavers (Carvalho *et al*, 2000; Oliveira-Costa *et al*, 2011).

Peckia lambens Wiedemann, 1830, sarcophagid with another geographical distribution Nearctic, Neotropical and Australasia, was appointed as specie of forensic importance in many such studies (Carvalho & Mello-Patiu 2008; Buenaventura *et al*, 2009; Vairo *et al*, 2011). It was recorded between Calyptrate flies associated with decomposition of domestic swine in Rio de Janeiro, in an area of Cerrado of the Distrito Federal and two Cerrado vegetation profiles in Uberlandia, Minas Gerais. In these studies, the presence of this dipterous was significant, being always among the most abundant species (Barros *et al,* 2008; Barbosa *et al,* 2009; Rosa *et al,* 2009).

Another important factor was its attraction for the housing decay phase, followed by the dry phase (Barros *et al,* 2008). The period in which the species used the substrate was dry (winter) and the profile of the dirty field (Rose *et al*, 2009). It has also been collected colonizing human cadavers, both adult and immature (Oliveira-Costa *et al,* 2011).

As holometabolous insects, *P. intermutans* and *P. lambens* undergo major structural, anatomical and morphological changes during its post-embryonic development. The changes that occur during intra-pupal development have been described as to some dipterous *Drosophila melanogaster* (Robertson 1936), *Musca domestica* and *Sarcophaga bullata* (Fraenkel & Bhaskaran 1973; Sivasubramanian & Biagi 1983), *Oestrus ovis* (Cepeda-Palacios & Scholl 2000), *Chrysomyia albiceps* (Pujol-Luz & Barros-Cordeiro 2012) and *Hermetia illucens* (Barros-Cordeiro *et al,* 2014).

Some authors reported the pupa length of time for *Peckia* genre: *P. intermutans* (Loureiro *et al*, 2005), *P. ingens* and *P. chrysostoma* (Ferraz 1985), *P. gulo* (Mendez & Pape 2002), *P. smarti* and *P. pallidipilosa* (Oliveira-da-Silva *et al*, 2006). As for studies related to description of the puparium, there is only a description of the puparium of *Peckia collusor*

(Mendonça *et al*, 2013) performed with the aid of scanning electron microscopy.

The objective of this study was to describe and to analyze events and morphological changes during the intra-pupal development of *P. intermutans* temperature of 23 ° C and relative humidity of 73% and *P. lambens* under three conditions controlled laboratory (21 °, 26 ° and 31 ° ± 1 ° C; 60 ± 10% RH and photoperiod 12:12). And also assess viability rates, larval and pupal and the sex ratio exposed the different conditions proposed.

2. MATERIAL AND METHODS

2.1 Collects adults and montage of the colony

The collection of wild insects was held at the Experimental Biology Station (BSE) in the Campus Darcy Ribeiro and Fazenda Agua Limpa (FAL) at the University of Brasilia (UNB) in a closed area with the use of the type Van Someren-Rydon traps. As bait, they were used shrimp, sardines and bovine liver 24 hours exposure at room temperature. Was offered to females, ground beef with 48 hours of exposure at room temperature as larviposition substrate for mounting colony. The larvae used both in the experiment with *P. intermutans* as in experiments with *P. lambens* came from a single colony assembled in the laboratory.

2.2 Fixation and dissection of pupae

After the establishment of the colony, 350 larvae of *P. intermutans* were kept in the laboratory at ambient conditions. The temperature and relative humidity (RH) were recorded throughout the experiment every 6 hours with the aid of a weather station, and 23 ° C and 73% RH, the mean values. And 1,500 larvae of *P. lambens* were placed in climate chambers (BOD) adjusted to temperatures 21, 26 and 31 ± 1 ° C; relative humidity of 60% ± 10% and photoperiod of 12L: 12D. After the process of pupariation five *P. intermutans* pupae were fixed every six hours until the emergence of the first adults and ten *P. lambens* pupae were established every 3 hours to complete the first 24 hours and thereafter every 6 hours until the emergence of the first adults. A total of 270 pupae of *P. intermutans* and 936 pupae of *P. lambens* were dissected in the experiments. All specimens were fixed in Carnoy's solution for 48 hours, then in 5% formic acid for 48 hours also; and they were then continuously kept in 70% ethanol. The specimens were dissected under a microscope stereoscopic Leica S8 APO and photographed in a stereomicroscope (Leica M205 C) coupled with camera and software for imaging (LAS V8).

2.3 Calculation of larval viability, pupal, sex ratio of P. lambens

To observe the larval viability, were used larvae maintained at controlled temperatures (21, 26, 31 ° C). In the first experiment, seven groups of 100 and a group of 41; the second three groups 100 and one of 101 and the third experiment, three groups of 100 and one of 58 neolarvas were kept in plastic containers, each containing ground beef 48h exposure to room temperature (2g/larva), the percentage of the formed pupae was observed every day. For the pupal viability, other pupae after the fasteners were used, registering the percentage of adults emerged and the sex ratio was calculated by dividing the number of females by the total number of individuals (male + female).

3. RESULTS

3.1 Pupariation

Mature larvae ceased feeding, reduced mobility and size and their cuticles have become opaque, pigmented and hardened, this process *P. intermutans* lasted a minimum duration of 36 hours and *P. lambens* had a minimum duration of 48 hours at 21 ° C, 34 hours at 26 ° C and 28 hours at 31 ° C.

3.2 Apolysis larval-pupal

Process which results in the formation of adult epidermis and their subsequent separation from the last larval cuticle which form the puparium, after completion of pupariation process. In *P. intermutans* lasted 2.3 ± 3.1 hours and had a minimum duration of 6 hours (Table 1) and *P. lambens* lasted 3.0 ± 4.7; 1.0 ± 1.4 and 0.9 ± 1.8 hours and had a minimum duration of 3 hours (Tables 2, 3 and 4).

3.3 Cryptocephalic pupa

In that event the pupa has no definite shape and it is impossible to externally observe the distinction of the head and thoracic appendages (Figs. 1A, 3A). This event in *P. intermutans* lasted 28.5 ± 20.0 hours and was completed at least 6 hours (Table 1). And in *P. lambens* lasted 8.4 ± 7.4; 4.3 ± 2.2 and 5.1 ± 2.2 hours and had minimum 3 hours (Tables 2, 3 and 4).

3.4 Phanerocephalic pupa

The extroversion of the head and thoracic appendages ends and begins the process of apolysis between pupa and adult (Figs. 1B, 3B). This event in *P. intermutans* lasted 29.7 ± 9.5 hours and was completed at least 30 hours (Table 1) and in *P. lambens* lasted 9.0 ± 3.1; 8.6 ± 4.4 and 9.7 ± 2.4 hours and had the same minimum duration of 3 hours (Tables 2,3 and 4).

3.5 Pharate adult

It was the longest event observed, the intra-puparial development extending from the end of the pupal-adult apolysis until the emergence of the adult and was divided into phases based on the color of the compound eyes: transparent eyes; yellow eyes, pink eyes and red eyes that corresponds to maturation of the adult (Figures 1C-D, 3C-D.): (I) transparent eyes, 42nd - 78th hour, lasting 62.3 ± 10.0 hours in *P. intermutans* and *P. lambens*, 9th - 66th and 9th - 12th hour, lasting 27.3 ± 20.8; 10.7 ± 1.5 and 11.0 ± 1.4 hours (Figures 2A, 4A, Tables 1-4), followed by differentiation of the head, chest, abdomen and chest appendages, mouthparts, antennae and abdominal spiracles ; (II) yellow eyes, 66th and 234th hours, lasting 143.0 ± 47.1 hours in *P. intermutans* and *P. lambens*, 15th - 174th; 12th - 114th and 12th - 96th hour, lasting 91.1 ± 46.4; 52.1 ± 28.6 and 46.1 ± 23.6 hours (Figures 2B, 4B, Tables 1-4), the bristles, the eyespots were observed and began pigmentation; (III) pink eyes, 240th – 270th hours, lasting from 253.8 ± 9.8 hours in *P. intermutans* and *P. lambens* 168th – 192th; 90th - 78th and 126th - 108th hour, lasting 180.9 ± 5.7; 106.5 ± 10.3 and 92.8 ± 7.2 hours (Figures 2C, 4C, Tables 1-4), pigmentation was increased in all parts of the body, it has also started sclerotization chest and the three stripes and stains family characteristics abdomen became apparent; (IV) red eyes, 252th its 318th hours, lasting 289.7 ± 19.2 hours in *P. intermutans* and *P. lambens*, $180^{th} - 234^{th}$; 108th - 126th and 96th - 114th hours, lasting from 212.7 ± 14.2; 120.0 ± 5.3 and 107.2 ± 5.9 hours (Figures 2D, 4D, Table 1-4), the adult was completely formed and pigmented.

3.6 The imagoes and adult emergence

The imagoes were fully formed and were observed within hours 252th in *P. intermutans*, 180th, 108th and 96th in *P. lambens*. And the emergence of the first specie adults occurred from the 318th hour, and the second from the 234th; 126th and 114th hours (Tables 1- 4).

3.7 Larval viability, pupal and sex ratio of P. lambens

Seven hundred forty-one larvae of the experiment at $21 °$ C, ninety died before starting the process of pupariation, resulting in 87.6% of larval viability. In the second experiment at $26 °$ C, of 401 larvae, 10 died a viability of 97 5% and at $31 °$ C was 358 larvae, 25 of which have not reached the pupariation process, succeeding in 93% larval viability and pupal viability was 86%, 98% and 87% (Table 5).

In the first experiment, 184 specimens emerged, 86 males and 98 females, representing a sex ratio of 0.53. In the second, emerged 118 individuals, 55 males and 63 females, a sex ratio of 0.53. In the last experiment, emerged flies 90 of which 32 were males and 58 were females, resulting in a sex ratio of 0.64 (Table 5). It was observed in all three experiments that males than females first emerged.

Table 1. Intra-puparial development in *Peckia intermutans.*

Stage	Event	Time of development (hour) Mean ± SE (Range)	Minimum duration (hours)	Sample size (n)
	Larval-pupal apolysis	2,3 ± 1,03 (00-06)	6	8
	Cryptocephalic pupa	28,5 ± 5,53 (06-54)	6	12
	Phanerocephalic pupa	29,7 ± 1,93 (12-48)	30	23
Pupa	Pharate			
	Transparent eyes	62,3 ± 2,42 (42-78)	24	16
	Yellowish eyes	143,0 ± 5,55 (66-234)	174	71
	Pinkish eyes	253,8 ± 2,95 (240-270)	12	10
	Reddish eyes	289,7 ± 2,69 (252-318)	66	50

Table 2. Intra-puparial development in *Peckia lambens* at 21 ° C.

Stage	Event	Development time (hours) Mean ± SD (Range)	Duration minimum (hours)	Size sample (n)
	Larval-pupal apolysis	3.0 ±4.7 (00-15)	3	17
	Cryptocephalic pupa	8.4 ±7.4 (03-24)	3	16
	Phanerocephalic pupa	9.0 ± 3.1(06-15)	3	12
	Pharate adult			
Pupa	Transparent eyes	27.3 ± 20.8 (09-66)	6	33
	Yellowish eyes	91.1 ± 46.4 (15-174)	153	257
	Pinkish eyes	180.9 ± 5.7 (168-192)	12	27
	Reddish eyes	212.7 ± 14.2 (180-234)	54	75

Table 3. Intra-puparial development in *Peckia lambens* at 26°C.

Stage	Event	Development time (hours) Mean ± SD (Range)	Duration minimum (hours)	Size sample (n)
	Larval-pupal apolysis	1.0 ±1.4 (00-03)	3	15
	Cryptocephalic pupa	4.3 ± 2.2 (03-09)	3	7
	Phanerocephalic pupa	8.6 ± 4.4 (06-21)	3	14
	Pharate adult			
Pupa	Transparent eyes	10.7 ± 1.5 (09-12)	3	16
	Yellowish eyes	52.1 ± 28.6 (12-114)	78	171
	Pinkish eyes	106.5 ± 10.3 (90-126)	18	24
	Reddish eyes	120.0 ± 5.3(108-126)	18	23

Table 4. Intra-puparial development in *Peckia lambens* at 31°C.

Stage	Event	Development time (hours) Mean ± SD (Range)	Duration minimum (hours)	Size sample (n)
	Larval-pupal apolysis	0.9 ±1.8 (00-06)	3	13
	Cryptocephalic pupa	5.1 ± 2.2 (03-09)	3	17
	Phanerocephalic pupa	9.7 ±2.4 (06-12)	3	9
	Pharate adult			
Pupa	Transparent eyes	11.0 ± 1.4 (09-12)	3	3
	Yellowish eyes	46.1 ± 23.6 (12-96)	66	140
	Pinkish eyes	92.8 ± 7.2 (78-108)	18	17
	Reddish eyes	107.2 ± 5.9 (96-114)	18	30

Table 5. Larval and pupal viability, sex ratio of *Peckia lambens* at different temperatures.

Temperature	Larval viability	Pupal viability	Sex ratio
21°C	87.6%	87.8%	0.53
26°C	97.5%	98.1%	0.53
31°C	93%	92.8%	0.64

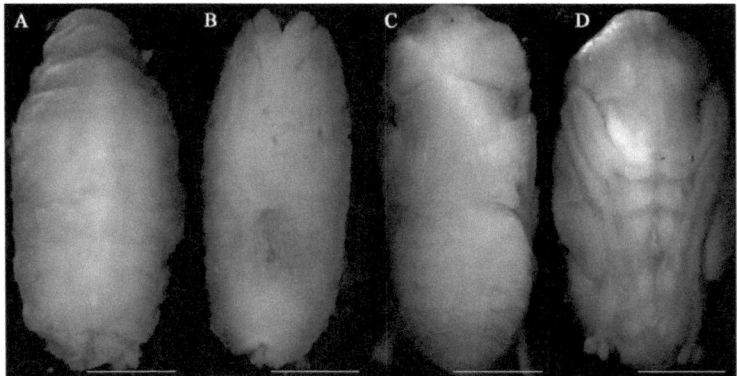

Figure 1. Sequence of the intra-puparial development of the *Peckia intermutans:* (A) Cryptocephalic pupa–dorsal view; (B) Phanerocephalic pupa –ventral view; (C, D) Pharate adult –dorsal and ventral view. **Scale bars:** 2 mm.

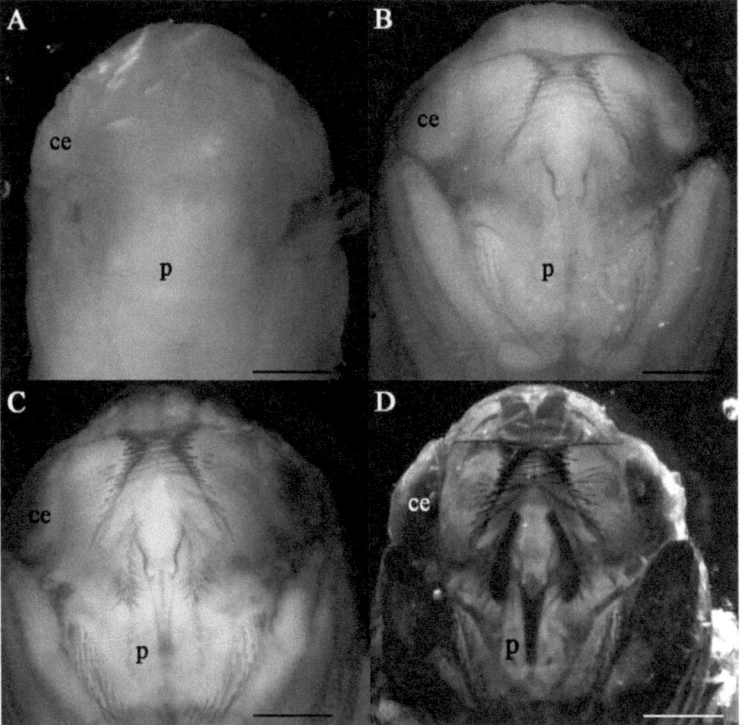

Figure 2. Pharate of the *Peckia intermutans* according to the color of the compound eyes: (A) transparent eyes; (B) yellowish eyes; (C) pinkish eyes (D) reddish eyes. **Abbreviations: ce,** compound eyes; **p,** proboscis. **Scale bars:** 2 mm.

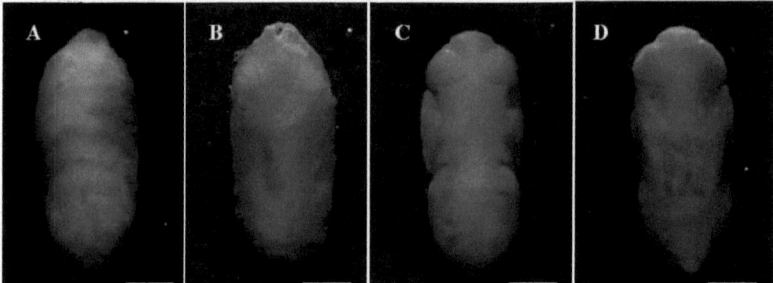

Figure 3. Sequence of the intra-puparial development in *Peckia lambens:* (A) Cryptocephalic pupa–dorsal view; (B) Phanerocephalic pupa–ventral view; (C, D) Pharate adult –dorsal and ventral view. **Scale bars:** 2 mm.

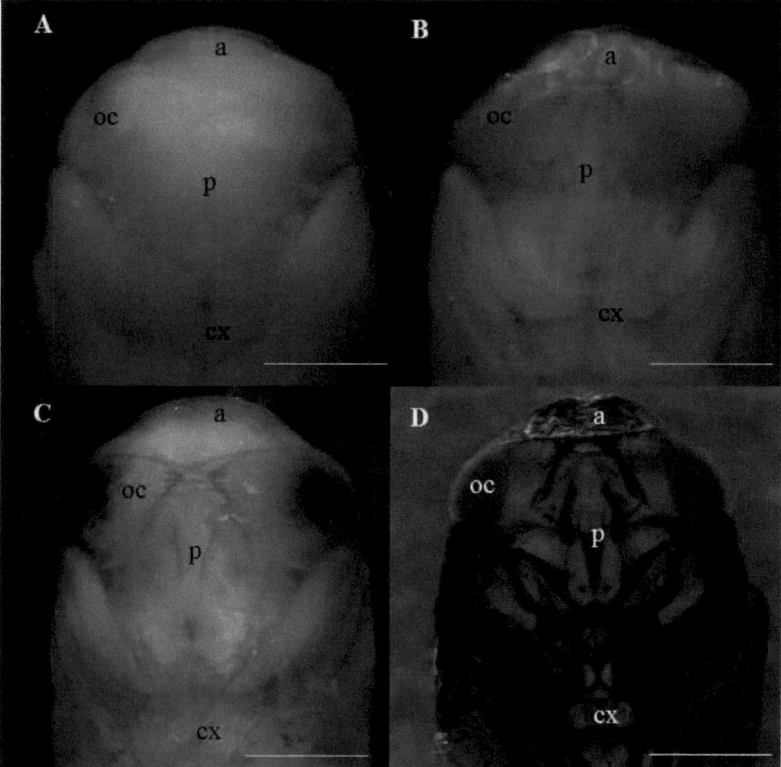

Figure 4. Pharate of the *Peckia lambens* according to the color of the compound eyes: (A) transparent eyes; (B) yellowish eyes; (C) pinkish eyes (D) reddish eyes. **Abbreviations: ce**, compound eyes; **p**, proboscis; **a**, antenna; **cx**, coxa. **Scale bars:** 1 mm.

4. DISCUSSION

In the two earlier studies on the morphological changes during intra-pupal development Sarcophagidae, from the same species *Sarcophaga bullata* Parker, 1916 (Fraenkel & Bhaskaran 1973; Sivasubramanian & Biagi 1983), there were no further details as shown in this work.

The apolysis larval-pupal of *S. bullata* a 24° C it lasted 20 hours, while at 25 ° C lasted 10 hours. In the present study, *P. intermutans* at 23°C had a minimum duration of 6 hours and *P. lambens* in the three temperatures (21, 26 and 31 ° C) the minimum duration was the same 3 hours. The hypotheses for these differences between species can be due to several factors not recorded in the previous works, such as the intervals at which occurred the fixings, offered the diet, the relative humidity and the said time, if the minimum or maximum. The pharate adult stage to 24 ° C occurred between 168th – 192th hours at 25 ° C took place between the 96th – 192th hours in *S. bullata*. In *P. intermutans* at 23 ° C occurred between 42nd – 252th hours and *P. lambens* occurred between 9th - 180th; 9th - 108th; 9th - 96th hours at temperatures 21, 26 and 31 ° C respectively.

Among muscoid flies, development time differs greatly between species. *S. bullata*, Fraenkel & Bhaskaran (1973) and Sivasubramanian & Biagi (1983) also found differences: at 24 ° C lasted 13.16 days and at 25 ° C, 11.25 days. In this study, 23 ° C, the emergence of adult *P. intermutans* occurred at 13.25 days. Using different diets, Loureiro et al. (2005), recorded the pupal time to *P. intermutans* at 27 ° C, 13.87 ± 0.51 in beef diet. However, pupa was recorded 18-19 days at 28 ° C and 28-30 days at 20 ° C in studies carried out by Bolaños & Jirón (1986). The wide variations observed for pupal time of the same species may be due to the time stamp told from the white pre-pupa. As it was not registered in the work throughout the methodology used, it cannot state the reasons for such differences.

The temperature seems to be a decisive factor for the development of flesh flies. For *P. trivittata* Curran, 1927 low temperatures (e.g. 16 ° C) tended to stop the development and dehydrate the adults, while development was complete at higher temperatures (Salviano *et al*, 1996). *P. chrysostoma* Wiedemann, 1830 and *P. ingens* Walker, 1849 held on temperatures 18, 25.9 and 27 ° C had lower development time at higher temperature (Ferraz 1995), corroborating the present study that showed similar results for *P. lambens* that at 21 ° C, development time was 9.75 days and 31 ° C was 4.5 days.

According to studies conducted with different species of forensic importance, abiotic factors influence the development time, the most relevant temperature than diets, differences in geographic origin and photoperiod (Milward-de-Azevedo *et al,* 1996; Villet *et al,* 2006; Nassu *et al,* 2014).

The 26 ° C average total viability of *P. lambens* was 97.8% value close to those obtained by Salviano et al. (1996), for *P. trivittata* (91.3% at 27 ° C). Similar results were also found for *P. chrysostoma* (95.5%) and *P. ingens* (89.5%) under a temperature of 25.9 ° C (Ferraz 1995).

The intra-pupal development *P. intermutans* and *P. lambens* the conditions set out in this work, with the specific differences, was similar to the species studied, especially the time used for the maturation of pharate adult which accounted for most of the period intra-puparial. With constant temperature, humidity and photoperiod, *P. lambens* showed some minor differences, however, persisted in the viability rate and sex ratio.

5. REFERENCES

Ames, C.; Turner, B. & Daniel, B. 2006. The use of mitochondrial cytochrome oxidase I gene (COI) to differentiate two UK blowfly species–*Calliphora vicina* and *Calliphora vomitoria*. **Forensic Science International 64:** 179–182.

Aspoas, B.R. 1991. Comparative micromorphology of third instar larvae and the breeding biology of some Afrotropical *Sarcophaga* (Diptera: Sarcophagidae). **Medical and Veterinary Entomology 5:** 437-445.

Banks, N. 1912. The structure of certain dipterous larvae with particular reference to those in human foods. **U.S. Department of Agriculture, Bureau of Entomology Technical Series (Bulletim) 22:** 44pp., illus.

Barros-Cordeiro, K.B., Báo, S.N., Pujol-Luz, J.R. 2014. Intra-puparial development of the Black soldier-fly *Hermetia illucens*. **Journal of Insect Science 14(83):** 1-10.

Barbosa, R.R, Mello-Patiu, C.A., Mello R.P., Queiroz M.M.C. 2009. New records of calyptrate dipterans (Fanniidae, Muscidae and Sarcophagidae) associated with the decomposition of domestic pigs in Brazil. **Memórias do Instituto Oswaldo Cruz 104:** 923-926.

Barros, R.M., Mello-Patiu, C.A., Pujol-Luz, J.R. 2008. Sarcophagidae (Insecta, Diptera) associados à decomposição de carcaças de *Sus scrofa* Linnaeus (Suidae) em área de Cerrado do Distrito Federal, Brasil. **Revista Brasileira de Entomologia 52(4):** 606-609.

Bohart, G.E., Gressitt, J.L. 1951. Filth-inhabiting flies Of Guam. **Bernice P.. Bishop Museum Bulletin 204:** 1-154.

Brown, B.V., Borkent, A., Cumming, J.M., Woodley, N.E., Zumbado, M. 2009. **Manual of Central American Diptera, vol. I**. Ottawa, NRC Research Press, 714p.

Buenaventura, E., Pape, T. 2013. Revision of the world genus *Peckia* Robineau-Desvoidy (Diptera: Sarcophagidae). **Zootaxa 3622(1):** 001-087.

Buenaventura, E., Camacho, G., García, A., Wolff, M. 2009. Sarcophagidae (Diptera) de importancia forense en Colombia: claves taxonômicas, notas sobre su biologia y distribuición. **Revista Colombiana de Entomología 35(2):** 189-196.

Cantrell, B.K. 1981. The immature stages of some Australian Sarcophaginae (Diptera: Sarcophagidae). **Journal of the Australian Entomological Society 20:** 237–248.

Cantrell, B.K. 1978. A new species of *Blaesoxipha* Loew from Australia (Diptera: Sarcophagidae). **Journal of the Australian Entomological Society 17:** 363-366.

Carvalho, C.J.B., Rafael, J. A., Couri, M.S., Silva, V.C. 2012. Diptera Linnaeus, 1758. *In*: **Insetos do Brasil: Diversidade e Taxonomia.** Ribeirão Preto: Holos. 810p.

Carvalho, C.J.B., Mello-Patiu, C.A. 2008. Key to the adults of the most common forensic species of Diptera in South America. **Revista Brasileira de Entomologia 52 (3):** 390-406.

Carvalho, L.M.L., Linhares, A.X. 2001. Seasonality of insect succession and pig carcass decomposition in a natural forest area in southeastern Brazil. **Journal Forensic Science 46:** 604-608.

Carvalho, L.M.L., Thyssen, P.J., Linhares, A.X., Palhares, F.A.B. 2000. A checklist of arthropods associated with pig carrion in human corpses in Southeastern Brazil. **Memórias do Instituto Oswaldo Cruz 95(1):** 135-138.

Cepeda-Palacios, R., Scholl, P.J. 2000. Intra-puparial development in *Oestrus ovis* (Diptera: Oestridae). **Journal of Medical Entomology 37:** 239-245.

Costa, C., Vanin, S. A. 1985. On the concepts of the "pre-pupa" with a special reference to the Coleoptera. **Revista Brasileira de Zoologia 2 (6):** 339-345.

Denlinger, D.L., Zdárek, J. 1994. Metamorphosis behavior of flies. **Annual Review of Entomology 39:** 243-266.

Draber-Monko, A., Malewski, T., Pomorski, J., Los, M., Slipinski, P. 2009. On the morphology and mitochondrial DNA Barcoding of the flesh fly *Sarcophaga (Liopygia) argyrostoma* (Robineau-Desvoidy, 1830) (Diptera: Sarcophagidae) – an important species in forensic entomology. **Annales Zoologici (Warszawa) 59 (4):** 465-493.

Fernandes. L.F., Pimenta, F.C., Fernandes, F.F. 2009. First report of human myiasis in Goiás State, Brazil: frenquency of different types of myiasis, their various etiological agents, and associated factors. **Journal of Parasitology 95(1):** 32-38.

Ferrar, P. 1979. The immature stages of dung-breeding muscoid flies in Australia, with notes on the species, and key to larvae and puparia. **Australian Journal of Zoology 73:** 1-106.

Ferraz, M.V. 1995. Larval and pupal periods of *Peckia chrysostoma* and *Adiscochaeta ingens* (Diptera: Sarcophagidae) reared under laboratory conditions. **Memórias do Instituto Oswaldo Cruz 90(5):** 611-614.

Fontoura, P., Oliveira-Costa, J., Ribeiro-Rocha, A. 2013. Identificação II – Imaturos de Diptera. *In:* **"Insetos peritos" A Entomologia Forense no Brasil.** 1 ed. Campinas, SP: Millennium Editora, 462p.

Fraenkel, G., Bhaskaran, G. 1973. Pupariation and pupation in Cyclorrhaphous flies (Diptera): terminology and interpretation. **Annals of the Entomological Society of America 66:** 418-422.

Fuller, M.E. 1938. On the biology and early stages of *Helicobia australis* (Sarcophaginae), a dipterous insect associated with grasshoppers.

Proceedings of the Linnean Society of New South Wales 63: 133-138.

Greene, C.T. 1925. The puparia and larvae of sarcophagid flies. Proceedings of the National Museum 66: 1-31.

Guimarães, J.H., Amorim, D.S. 2006. Diptera. *In:* Insetos Imaturos: Metamorfose e Identificação. Ribeirão Preto: Holos Editora, 249p.

Guimarães J.H, Papavero N. 1999. Myiasis in man and animals in the Neotropical regian: bibliography database. São Paulo, Plêiade / FAPESP. 308p.

Hagman, M., Pape, T., Schulte, R. 2005. Flesh fly myiasis (Diptera, Sarcophagidae) in peruvian poison frogs genus *Epipedobates* (Anura, Dendrobatidae). Phyllomedusa 4(1): 69-73.

Hernández, J.V., Osborn, F., Herrera, B., Liendo-Barandiaran, C.V., Perozo, J., Velásquez, D. 2009. Parasitoides larva-pupa de *Hylesia metabus* (Lepidoptera: Saturniidae) en la región Nororiental de Venezuela: um caso de controle biológico natural. Neotropical Entomology 38(2):243-250.

Hinton, H. E. 1973. Neglected phases in metamorphosis: a reply to V. B. Wigglesworth. Journal of Entomology (A) 48 (1): 57-68.

Hinton, H.E. 1971. Some neglected phases in metamorphosis. Proceedings of Royal Entomological Society of London 35: 55-64.

Hinton, H.E. 1946. Concealed phases in the metamorphosis of insects. Nature 157: 552-553.

Ishijima, H. 1967. Revision of the third stage larvae of synanthropic flies of Japan (Diptera: Anthomyiidae, Muscidae, Calliphoridae and Sarcophagidae). Japanese Journal of Sanitary Zoology 18(2/3): 46-100.

Jenkin, P.M., Hinton, H.E. 1966. Apolysis in arthropod moulting cycles. Nature 5051: 871.

Jirón, L.F., Bolaños, R. 1986. Biology and larval morphology by scanning electron microscopy of *Pattonella intermutans* Walker (Diptera, Sarcophagidae). **Revista Brasileira de Entomologia 30(1): 27-30.**

Kamal, A.S. 1958. Comparative study of thirteen species of sarcosaprophagus Calliphoridae and Sarcophagidae (Diptera) 1. Bionomics. **Annals Entomological Society of America 51: 261-271.**

Kano, R., Lopes, H.S. 1971. A new genus, *Johnsonimima,* and two new species belonging to this genus from the Solomon Islands (Diptera, Sarcophagidae). **Pacific Insects, 13 (3/4): 597-602.**

Knipling, E.F. 1936. A comparative study of the first instar larvae of the genus *Sarcophaga* (Calliphoridae: Diptera) with notes on the biology. **The Journal of Parasitology 22(5): 417-454.**

Ledo, R.M.D., Barros, R.M., Pujol-Luz, J.R. 2012. Sarcophagidae and Calliphoridae related to *Rhinella schneideri* (Anura, Bufonidae), *Bothrops moojeni* (Reptilia, Serpentes) and *Mabuya frenata* (Reptilia, Lacertilia) carcasses in Brasília, Brasil. **Revista Brasileira de Entomologia 56(3): 377-380.**

Leite, A.C.R., Lopes, H.S. 1989. Scanning electron microscopy of the first instar larvae of *Sarcodexia lambens* e *Peckia chrysostoma* (Diptera: Sarcophagidae). **Memórias do Instituto Oswaldo Cruz 84: 303-307.**

Lopes, H.S., Leite, A.C.R. 1989. Morphology of the egg of *Sarcodexia lambens* (Diptera: Sarcophagidae). **Memórias do Instituto Oswaldo Cruz 84 (4): 497-500.**

Lopes, H.S., Leite, A.C. 1986. Studies on some features of the first instar larvae of *Oxysarcodexia* (Diptera, Sarcophagidae) based on scanning electron microscope observations. **Revista Brasileira de Biologia 46(4): 741-746.**

Lopes, H.S. 1985. Descriptions of six new species of *Retrocitomyia* Lopes (Diptera, Sarcophagidae). **Boletim do Museu Nacional 309:** 1-8.

Lopes, H.S. 1983. On *Notochaetomima* (Diptera, Sarcophagidae) with descriptions of four new species, one of them living on *Beltela sp* (Mollusca, Gastropoda). **Revista Brasileira de Entomologia 27(3/4):** 259-266.

Lopes, H.S. 1982a. The importance of the mandible and clypeal arch of the first larvae in the classification of the Sarcophagidae (Diptera). **Revista Brasileira de Entomologia, 26(3/4):** 293-326.

Lopes, H.S. 1976. On the genus *Cuculomyia* Roback (Diptera, Sarcophagidae). **Revista Brasileira de Biologia, 36(4):** 745-757.

Lopes, H.S. 1975. Notes on *Amblycoryphenes* Townsend, 1918 (Diptera, Sarcophagidae, Tephromyini). **Revista Brasileira de Biologia, 35(2):** 265-271.

Lopes, H. S. 1971. Notes on *Emblemasoma* and *Pessoamyia* (Diptera, Sarcophagidae). **Revista Brasileira de Biologia 31(1):** 89-97.

Lopes, H. S. 1968. Sobre uma espécie nova de "Dexosarcophaga" Townsend, 1917 (Dipt., Sarcophagidae) cujas larvas vivem em ninho de "Camponotus" (Hymenoptera, Formicidae). **Revista Brasileira de Biologia 28 (4):** 521-523.

Lopes, H.S. 1954. Contribuição ao conhecimento do gênero *Sarcophagula* Wulp, 1887 (Diptera-Sarcophagidae). **Memórias do Instituto Oswaldo Cruz, 52 (3/4):** 587-602.

Lopes, H.S., Vogelsang, E.G. 1953. *Notochaeta bufonivora* n.sp., parasita de *Bufo granulosus spix* em Venezuela (Diptera Sarcophagidae). **Anais da Academia Brasileira de Ciências 25(2):** 139-143.

Lopes, H.S. 1946. Novos sarcofagídeos neotrópicos representados na coleção "Imperial Institute of Entomology" (Diptera, Sarcophagidae). **Revista Brasileira de Biologia 6(1):** 117-131.

Lopes, H.S. 1945. Contribuição ao conhecimento das espécies do gênero *Notochaeta* Aldrich, 1916. (Diptera-Sarcophagidae). **Memórias do Instituto Oswaldo Cruz 42(3):** 503–550.

Lopes, H.S. 1943. Contribuição ao conhecimento das larvas dos Sarcophagidae com especial referência ao esqueleto cefálico (Diptera). **Memórias Instituto Oswaldo Cruz 38 (2):** 127–163.

Lopes, H.S. 1935. Sobre duas espécies de *Sarcophaga* cujas larvas são predadoras (Dipt. Sarcophagidae). **Revista de Entomologia 5 (4):** 470- 479.

Loureiro, M.S., Oliveira, V.C., d'Almeida, J.M. 2005. Desenvolvimento pós-embrionário de *Pattonella intermutans* (Thomson) (Diptera: Sarcophagidae) em diferentes dietas. **Revista Brasileira de Entomologia 49(1):** 127-129.

Marchiori, C.H., Leles, A.S., Carvalho, S.A., Rodrigues, R.F. 2007. Parasitoides de dípteros muscoides coletados no Matadouro Alvorada em Itumbiara, sul de Goiás, Brasil. **Revista brasileira de parasitologia veterinária, 16(4):** 235-237.

Marchiori, C.H., Pereira, L.A., Filho, O.M.S. 2003. Primeiro relato do parasitoide *Pachycrepoideus vindemiae* Rondani (Hymenoptera: Pteromalidae) parasitando pupas de *Sarcodexia lambens* Wiedemann (Diptera: Sarcophagidae) no Brasil. **Ciência Rural 33:** 173-175.

Mendez, J., Pape, T. 2002. Biology and immatures stages of *Peckia gulo* (Fabricius, 1805) (Diptera: Sarcophagidae). **Studia Dipterologica 9:** 371-374.

Mendonça, P.M., Cortinhas, L.B., Santos-Mallet, J.R., Queiroz, M.M.C. 2013. Ultrastructure of immature stages of *Peckia (Euboetcheria) collusor* (Diptera: Sarcophagidae). **Acta Tropica 128(3):** 522-527.

Mello-Patiu, C.A.; Paseto, M.L.; Faria, L.S.; Mendes, J. & Linhares, A.X. 2014. Sarcophagid flies (Insecta, Diptera) from pig carcasses in

Minas Gerais, Brazil, with nine new records from the Cerrado, a threatened Neotropical biome. **Revista Brasileira de Entomologia 58 (2):** 142-146.

Milward-de-Azevedo, E.V.; Carraro, V.M.; Martins, C.; Moreira, O.I.; Cruz, M. & Serafin, I. 1996. Desenvolvimento pós-embrionário de *Chrysomyia megacephala* (Fabricius) (Diptera: Calliphoridae) em diferentes temperaturas, sob condições experimentais. Parte 1. 1996. **Arquivos de Biologia e Tecnologia 39 (4):** 793-798.

Miranda, G.H.B.; Costa, K.A. & Pujol-Luz, J.R. 2013. Vestígios entomológicos. p. 125-150 In: **Locais de Crime.** Velho, J.A.; Costa, K.A. & Damasceno, C.T.M. (Org.). Millenium Editora. Campinas. Xviii+574.

Moretti, T.C., Ribeiro O.B., Thyssen, P. J., Solis, D.R. 2008. Insects on decomposing carcasses of small rodents in a secondary forest in Southeastern Brazil. **European Journal of Entomology 105:** 691–696.

Nassu, M.P.; Thyssen, P.J. & Linhares, A.X. 2014. Developmental rate of immatures of two fly species of forensic importance: *Sarcophaga (Liopygia) ruficornis* and *Microcerella halli* (Diptera: Sarcophagidae). **Parasitology Research 113:** 217-222.

Newhouse, V.F., Walker, D.W., James, M.T. 1955. The immature stages of *Sarcophaga cooleyi, S. bullata,* and *S. shermani* (Diptera: Sarcophagidae). **Journal of the Washington Academy of Sciences 45(1):** 15-20.

Oliveira-Costa, J., Queiroz, M.M.C., Azevedo, A.P., Santana, D.O. 2011. Dípteros de interesse forense no Brasil. In: Oliveira-Costa, J. (Ed.), **Entomologia forense: quando os insetos são os vestígios.** 3 ed. Campinas, SP: Millennium Editora, pp. 87-130.

Oliveira-da-Silva, A., Ale-Rocha, R., Rafael, J.A. 2006. Bionomia dos estágios imaturos de duas espécies de *Peckia* (Diptera, Sarcophagidae) em suíno em decomposição em área de floresta no norte do Brasil. **Revista Brasileira de Entomologia 50(4):** 524-527.

Oliveira, V.C., D'Almeida, J.M., Paes, M.J., Sanavria, A. 2002a. Population dynamics of Calyptrate Diptera (Muscidae and Sarcophagidae) at the Rio-Zoo foundation, Rio de Janeiro, RJ, Brazil. **Brazilian Journal of Biology 62(2):** 191-196.

Oliveira, V.C., Mello, R.P., Santos, R.F.S. 2002b. Bionomic aspects of *Pantonella intermutans* (Thomson, 1869) (Diptera, Sarcophagidae) under laboratory conditions. **Brazilian Archives of Biology and Technology 45(4):** 473-477.

Pape, T., Blagoderov, V., Mostovski, M.B. 2011. Order Diptera Linnaeus, 1758. *In*: Zhang, Z.-Q. (Ed.) **Animal biodiversity: an outline of higher-level classification and survey of taxonomic richness. Zootaxa, 3148**, 237p.

Pape, T., Dahlem, G.A 2010. Sarcophagidae (flesh flies), p. 1313-1335. *In:* B.V. Brown, A. Borkent, J.M. Cumming, N.E. Woodley, & M. Zumbado (eds.). **Manual of Central American Diptera, vol**.II. Ottawa, NRC Research Press, 728p.

Pujol-Luz, J.R., Barros-Cordeiro, K.B. 2012. Intra-puparial development of the females of *Chrysomya albiceps* (Wiedemann) (Diptera, Calliphoridae). **Revista Brasileira de Entomologia 56(3):** 269-272.

Pujol-Luz, J.R.; Marques, H.; Rodrigues. A.U.; Rafael, J.A.; Santana, F.H.A.; Chaves, L. & Constantino, R. 2006. A forensic entomology case from the Amazon rain forest. **Jounal of Forensic Science 51**: 1-3.

Robertson, C.W. 1936. The metamorphosis of *Drosophila melanogaster*, including an accurately timed logical changes account of the principal morphological changes. **Journal of Morphology 59 (2):** 351-399.

Rocha, F.R., Mendes, J. 1996. Pupation of *Dermatobia hominis* (L. Jr., 1781) (Diptera: Cuterebridae) associated with *Sarcodexia lambens* (Wiedmann, 1830) (Diptera: Sarcophagidae). **Memórias do Instituto Oswaldo Cruz 91 (3):** 299-300.

Romera, E.; Arnaldos, M.I.; García, M.D. & Gonzáles-Mora, D. 2003. Los Sarcophagidae (Insecta, Diptera) de un ecosistema cadavérico en el sureste de la Península Ibérica. **Anales de Biologia 25:** 49-63.

Root, F.M. 1923. Notes on larval characters in the genus *Sarcophaga*. **The Journal of Parasitology 9(4):** 227-229.

Rosa, T.A., Bavata, M.L.Y., Souza, C.M., Sousa, D., Mello-Patiu, C.A., Mendes, J. 2009. Dípteros de interesse forense em dois perfis de vegetação de Cerrado em Uberlândia, MG. **Neotropical Entomology 38(6):** 859-866.

Saloña Bordas, M.I., González-Mora, D. 2005. Primera cita de *Liosarcophaga aegyptica* (Salem, 1935) (Diptera; Sarcophagidae) de la Península Ibérica, com descripción de sus fases larvárias II y III, pupario y adultos. **Boletín Sociedad Entomológica Aragonesa 36:** 251-255.

Salviano, R.J.B.; Mello, R.P.; Beck, L.C.N.H. & D'Almeida, J.M. 1996. Aspectos bionômicos de *Squamatoides trivittatus* (Diptera, Sarcophagidae) sob condições de laboratório. **Memórias do Instituto Oswaldo Cruz 91(2):** 249-254.

Shewell, G.E. 1987. Sarcophagidae, p.1159-1186. *In:* J.F. McAlpine, B.V. Peterson, G.E., Shewell, H.J. Teskey, J.R. Vockeroth & D.M. Wood (eds.). **Manual of Neartic Diptera, vol II.** Agriculture Canada Monograph 28, 675-1332.

Silva, C.G., Cruz, G.C., Filho, C.L., Araújo, W.J.S., Santos, L.E.A., Siqueira, T.S. 2012. Ocorrência de *Brachymeria podagrica* em pupas de sarcofagídeos no Estado do Maranhão. **Revista Trópica – Ciências Agrárias e Biológicas 6(2):** 89-92.

Singh, D; Garg; Rashmi. & Wadhawan. 2012. Ultramorphological characteristic of immature stages of a forensically important fly *Parasarcophaga ruficornis* (Fabricius) (Diptera: Sarcophagidae). **Parasitology Research 110:** 821-831.

Sivasubramanian, P., Biagi, M. 1983. Scientific note morphology of the pupal stages of the fleshfly, *Sarcophaga bullata* (Parker) (Diptera: Sarcophagidae). International **Journal Insect Morphology & Embryology 12 (5/6):** 355-359.

Stegmaier, C.E.Jr. 1972. Notes on some Sarcophagidae (Diptera) reared from snails (Mollusca) in Florida. **The Florida Entomologist 55(4):** 237-243.

Sukontason, K.L; Piangjai, S; Bunchu, N; Chaiwong, T; Sripakdee, D; Boonsriwong, W; Vogtsberger, R.C. & Sukontason, K. 2006. Surface ultrastructure of the puparia of the blow fly, *Lucilia cuprina* (Diptera: Calliphoridae), and flesh fly, *Liosarcophaga dux* (Diptera: Sarcophagidae). **The Journal Parasitology Research 98:** 482-487.

Vairo, K.P. 2011. Sarcophagidae (Diptera) de potencial interesse forense de Curitiba, Paraná: chave pictórica para as espécies e morfologia dos estágios imaturos de *Sarcodexia lambens* (Wiedemann). Dissertação de Mestrado, Universidade Federal do Paraná, Brasil.

Vairo, K. P., Mello-Patiu, C.A., Carvalho, C.J.B. 2011. Pictorial identification key for species of Sarcophagidae (Diptera) of potential forensic importance in Southern Brazil. **Revista Brasileira de Entomologia 55(3):** 333-347.

Villet, M.H.; Mackenzie, B. & Muller, W.J. 2006. Larval development of carrion-breeding flesh fly, *Sarcophaga (Liosarcophaga) tibialis* Macquart (Diptera: Sarcophagidae), at constant temperatures. **African Entomology 14 (2):** 357-366.

Wigglesworth, V.B. 1973. The significance of "apolysis" in the moulting of insects. **Journal Entomology (A) 47 (2):** 141-149.

Zumpt F. 1965. **Myiasis in man and animals in the Old World.** Butterworths, London, 267p.

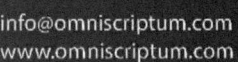

Printed by Books on Demand GmbH, Norderstedt / Germany